Berenice Blackburn was born in the nineteen fifties, just in time – as she says herself – 'to enjoy a solid grammar school education.' At that time girls were hardly encouraged to take up scientific pursuits, but Berenice did so with enthusiasm. Afterwards, university and then the world of work.

In all her working career, Berenice worked for only two companies. She never rose as high as she hoped. 'The glass ceiling was very thick then,' she says, 'and still exists, I believe.'

Asked to mention her finest achievement, Miss Blackburn points to her vegetable garden, which she claims not only feeds her organic produce but helps to keep her fit.

THE MAKING OF SOAP

Berenice Blackburn

THE MAKING OF SOAP

EMMA
STERN
PUBLISHING

An Emma Stern Publication

Copyright © Berenice Blackburn 2016

The right of Berenice Blackburn to be identified as author of this work
has been asserted in accordance with sections 77 and 78 of the
Copyright, Designs and Patents Act 1988.

A CIP catalogue record for this title is available from the British Library.

ISBN: 978-1-911224-09-9

Published in 2016

Emma Stern Publishing
107 Fleet Street
London
EC4A 2AB

www.emmastern.com
www.facebook.com/emmasternpublishing
Email: editorial@emmastern.com
Email: marketing@emmastern.com

Printed in Great Britain

worked in industry for several years and rose steadily up the ladder of preferment but, alas, then came up against the inevitable glass ceiling that stops female scientists from reaching the very highest grades.

I wish to dedicate this book to the few women who did actually make the long ascent and were allowed to influence policies at board level.

RAW MATERIALS USED IN SOAP MAKING

Soap is ordinarily thought of as the common cleansing agent well known to everyone. In a general and strictly chemical sense this term is applied to the salts of the non-volatile fatty acids. These salts are not only those formed by the alkali metals, sodium and potassium, but also those formed by the heavy metals and alkaline earths. Thus we have the insoluble soaps of lime and magnesia formed when we attempt to wash in "hard water"; again aluminum soaps are used extensively in polishing materials and to thicken lubricating oils; ammonia or "benzine" soaps are employed among the dry cleaners. Commonly, however, when we speak of soap we limit it to the sodium or potassium salt of a higher fatty acid.

It is very generally known that soap is made by combining a fat or oil with a water solution of sodium hydroxide (caustic soda lye), or potassium hydroxide (caustic potash). Sodium soaps are always harder than potassium soaps, provided the same fat or oil is used in both cases.

The detergent properties of soap are due to the fact that it acts as an alkali regulator, that is, when water comes into contact with soap, it undergoes what is called hydrolytic dissociation. This means that it is broken down by water into other substances. Just what these substances are is subject to controversy, though it is presumed caustic alkali and the acid alkali salt of the fatty acids are formed.

OILS AND FATS

There is no sharp distinction between fat and oil. By "oil" the layman has the impression of a liquid which at warm temperature will flow as a slippery, lubricating, viscous fluid; by "fat" he understands a greasy, solid substance unctuous to the touch. It thus becomes necessary to differentiate the oils and fats used in the manufacture of soap.

Soap is the alkali salt of a fatty acid, the oil or fat from which soap is made must have as a constituent part, these fatty acids. Hydrocarbon oils or paraffins, included in the term 'oil,' are thus useless in the process of soap-making, as far as entering into chemical combination with the caustic alkalis is concerned. The oils and fats which form soap are those which are a combination of fatty acids and glycerine, the glycerine being obtained as a by-product to the soap-making industry.

NATURE OF A FAT OR OIL USED IN SOAP MANUFACTURE

Glycerine, being a trihydric alcohol, has three atoms of hydrogen which are replaceable by three univalent radicals of the higher members of the fatty acids, e. g.,

OH
OR
$C_3 H_5$
OH
$+ 3 ROH = C_3 H_5$
OR
$+ 3 H_2O$
OH
OR
Glycerine plus 3 Fatty Alcohols equals Fat or Oil plus 3 Water.

Thus three fatty acid radicals combine with one glycerine to form a true neutral oil or fat which are called triglycerides. The fatty acids which most commonly enter into combination of fats and oils are lauric, myristic, palmitic, stearic and oleic acids and form the neutral oils or triglycerides derived from these, e. g., stearin, palmatin, olein. Mono and diglycerides are also present in fats.

SAPONIFICATION DEFINED.

When a fat or oil enters into chemical combination with one of the caustic hydrates in the presence of water, the process is called "saponification" and the new compounds formed are soap and glycerine, thus:

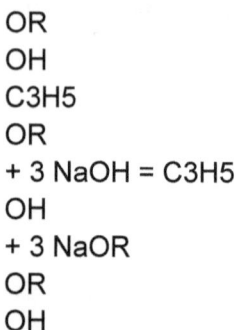

OR
OH
C_3H_5
OR
+ 3 NaOH = C_3H_5
OH
+ 3 NaOR
OR
OH

Fat or Oil plus 3 Sodium Hydrate equals Glycerine plus 3 Soap.

It is by this reaction almost all of the soap used today is made.

There are also other means of saponification, as, the hydrolysis of an oil or fat by the action of hydrochloric or sulphuric acid, by autoclave and by ferments or enzymes. By these latter processes the fatty acids and glycerine are obtained directly, no soap being formed.

FATS AND OILS USED IN SOAP MANUFACTURE

The various and most important oils and fats used in the manufacture of soap are, tallow, coconut oil, palm oil, olive oil, poppy oil, sesame oil, soya bean oil, cotton-seed oil, corn oil and the various greases. Besides these the fatty acids, stearic, red oil (oleic acid) are more or less extensively used. These oils, fats and fatty acids, while they vary from time to time and to some extent as to their colour, odour and consistency, can readily be distinguished by various physical and chemical constants.

Much can be learned by someone, who through continued acquaintance with these oils has become thoroughly familiarized with the indications of a good or bad oil, by taste, smell, feel and appearance. It is, however, not well for the manufacturer in purchasing to depend entirely upon these simpler tests. Since he is interested in the yield of glycerine, the largest possible yield of soap per pound of soap stock and the general body and appearance of the finished product, the chemical tests upon which these depend should be made. Those especially important are the acid value, percentage unsaponifiable matter and titre test.

A short description of the various oils and fats mentioned is sufficient for their use in the soap industry.

Tallow is the name given to the fat extracted from the solid fat or "suet" of cattle, sheep or horses. The quality varies greatly, depending upon the seasons of the year, the food and age of the animal and the method of rendering. It comes to the market under the distinction of edible and inedible, a further distinction being made in commerce as beef tallow, mutton tallow or horse tallow. The better quality is white and bleaches whiter upon exposure to air and light, though it usually has a yellowish tint, a well defined grain and a clean odour. It consists chiefly of stearin, palmitin and olein. Tallow is by far the most extensively used and important fat in the making of soap.

In the manufacture of soaps for toilet purposes, it is usually necessary to produce as white a product as possible. In order to do this it is often necessary to bleach the tallow before saponification. The method usually employed is the Fuller's Earth process.

FULLER'S EARTH PROCESS FOR BLEACHING TALLOW

From one to two tons of tallow are melted out into the bleaching tank. This tank is jacketed, made of iron and provided with a good agitator designed to stir up sediment or a coil provided with tangential downward opening perforations and a draw-off cock at the bottom. The coil is the far simpler arrangement, cleaner and less likely to cause trouble. By this arrangement compressed air which is really essential in the utilization of the press (see later) is utilized for agitation. A dry steam coil in an ordinary tank may be employed in place of a jacketed tank, which lessens the cost of installation.

The tallow in the bleaching tank is heated to 180° F. (82° C.) and ten pounds of dry salt per ton of fat used added and thoroughly mixed by agitation. This addition coagulates any albumen and dehydrates the fat. The whole mass is allowed to settle over night where possible, or for at least five hours. Any brine which has separated is drawn off from the bottom and the temperature of the fat is then raised to 160° F. (71° C).

Five per cent. of the weight of the tallow operated upon, of dry Fuller's earth is now added and the whole mass agitated from twenty to thirty minutes.

The new bleached fat, containing the Fuller's earth is pumped directly to a previously heated filter press and the issuing clear oil run directly to the soap kettle.

One of the difficulties experienced in the process is the heating of the press to a temperature sufficient to prevent solidification of the fat without raising the press to too great a temperature. To overcome this the first plate is heated by wet steam. Air delivered from a blower and heated by passage through a series of coils raised to a high temperature by external application of heat (super-heated steam) is then substituted for the steam. The moisture produced by the condensation of the steam is vaporized by the hot air and carried on gradually to each succeeding plate where it again condenses and vaporizes. In this way the small quantity of water is carried through the entire press, raising its temperature to 80°-100° C. This temperature is subsequently maintained by the passage of hot air. By this method of heating the poor conductivity of hot air is overcome through the intermediary action of a liquid vapour and the latent heat of steam is utilized to obtain the initial rise in temperature. To heat a small press economically where conditions are such that a large output is not required the entire press may be encased in a small wooden house which can be heated by steam coils. The cake in the press is heated for some time after the filtration is complete to assist drainage. After such treatment the cake should contain approximately 15 per cent. fat and 25 per cent. water. The cake is now removed from the press and transferred to a small tank where it is treated with sufficient caustic soda to convert the fat content into soap.

Saturated brine is then added to salt out the soap, the Fuller's earth is allowed to settle to the bottom of the tank and the soap which solidifies after a short time is skimmed off to be used in a cheap soap where colour is not important. The liquor underneath may also be run off without disturbing the sediment to be used in graining a similar cheap soap. The waste Fuller's earth contains about 0.1 to 0.3 per cent. of fat.

METHOD FOR FURTHER IMPROVEMENT OF COLOUR

A further improvement of the colour of the tallow may be obtained by freeing it from a portion of its free fatty acids, either with or without previous Fuller's earth bleaching.

To carry out this process the melted fat is allowed to settle and as much water as possible taken off. The temperature is then raised to 160° F. with dry steam and enough saturated solution of soda ash added to remove 0.5 per cent. of the free fatty acids, while agitating the mass thoroughly mechanically or by air. The agitation is continued ten minutes, the whole allowed to settle for two hours and the foots drawn off. The soap thus formed entangles a large proportion of the impurities of the fat.

VEGETABLE OILS

Coconut Oil, as the name implies, is obtained from the fruit of the coconut palm. This oil is a solid, white fat at ordinary temperature, having a bland taste and a characteristic odour. It is rarely adulterated and is very readily saponified. In recent years the price of this oil has increased materially because coconut oil is now being used extensively for edible purposes, especially in the making of oleomargarine. Present indications are that shortly very little high grade oil will be employed for soap manufacture since the demand for oleomargarine is constantly increasing and since new methods of refining the oil for this purpose are constantly being devised.

The oil is found in the market under three different grades: (1) Cochin coconut oil, the choicest oil comes from Cochin (Malabar). This product, being more carefully cultivated and refined than the other grades, is whiter, cleaner and contains a smaller percentage of free acid. (2) Ceylon coconut oil, coming chiefly from Ceylon, is usually of a yellowish tint and more acrid in odour than Cochin oil. (3) Continental coconut oil (Copra, Freudenberg) is obtained from the dried kernels, the copra, which are shipped to Europe in large quantities, where the oil is extracted. These dried kernels yield 60 to 70 per cent oil. This product is generally superior to the Ceylon oil and may be used as a very satisfactory substitute for Cochin oil, in soap manufacture, provided it is low in free acid and of good colour. The writer has employed it satisfactorily in the whitest and finest of toilet soaps without being able to distinguish any disadvantage to the Cochin oil. Since continental oil is usually cheaper than Cochin oil, it is advisable to use it, as occasion permits.

Coconut oil is used extensively in toilet soap making, usually in connection with tallow. When used alone the soap made from this oil forms a lather, which comes up rapidly but which is fluffy and dries quickly. A pure tallow soap lathers very much slower but produces a more lasting lather. Thus the advantage of using coconut oil in soap is seen. It is further used in making a coconut oil soap by the cold process also for "fake" or filled soaps. The fatty acid content readily starts the saponification which takes place easily with a strong lye (25°-35° B.). Where large quantities of the oil are saponified care must be exercised as the soap formed suddenly rises or puffs up and may boil over. coconut oil soap takes up large quantities of water, cases having been cited where a 500 per cent. yield has been obtained. This water of course dries out again upon exposure to the air. The soap is harsh to the skin, develops rancidity and darkens readily.

Palm Kernel Oil, which is obtained from the kernels of the palm tree of West Africa, is used in soap making to replace coconut oil where the lower price warrants its use. It resembles coconut oil in respect to saponification and in forming a very similar soap. Kernel oil is white in colour, has a pleasant nutty odour when fresh, but rapidly develops free acid, which runs to a high percentage.

Palm Oil is produced from the fruit of the several species of the palm tree on the western coast of Africa generally, but also in the Philippines. The fresh oil has a deep orange yellow tint not destroyed by saponification, a sweetish taste and an odour of orris root or violet which is also imparted to soap made from it. The methods by which the natives obtain the oil are crude and depend upon a fermentation, or putrefaction. Large quantities are said to be wasted because of this fact. The oil contains impurities in the form of fermentable fibre and albuminous matter, and consequently develops free fatty acid rapidly. Samples tested for free acid have been found to have hydrolized completely and one seldom obtains an oil with low acid content. Because of this high percentage of free fatty acid, the glycerine yield is small, though the neutral oil should produce approximately 12 per cent. glycerine. Some writers claim that glycerine exists in the free state in palm oil. The writer has in her work washed large quantities of the oil and analysed the wash water for glycerine. The results showed that the amount present did not merit its recovery. Most soap makers do not attempt to recover the glycerine from this oil, when used alone for soap manufacture.

There are several grades of palm oil in commerce, but in toilet soap making it is advisable to utilize only Lagos palm oil, which is the best grade. Where it is desired to maintain the colour of the soap this oil produces, a small quantity of the lower or brass grade of palm oil may be used, as the soap made from the better grades of oil gradually bleaches and loses its orange yellow colour.

Palm oil produces a crumbly soap which cannot readily be milled and is termed "short." When used with tallow and coconut oil, or 20 to 25 per cent. coconut oil, it produces a very satisfactory toilet soap. In the saponification of palm oil it is not advisable to combine it with tallow in the kettle, as the two do not readily mix.

Since the finished soap has conveyed to it the orange colour of the oil, the oil is bleached before saponification. Oxidation readily destroys the colouring matter, while heat and light assist materially. The methods generally employed are by the use of oxygen developed by bichromates and hydrochloric acid and the direct bleaching through the agency of the oxygen of the air.

CHROME BLEACHING OF PALM OIL

The chrome process of bleaching palm oil is more rapid and the oxygen thus derived being more active will bleach oils which air alone cannot. It depends upon the reaction:

$$Na_2Cr_2O_7 + 8HCl = Cr_2Cl_6 + 2NaCl + 7O.$$

in which the oxygen is the active principle. In practice it is found necessary to use an excess of acid over that theoretically indicated.

For the best results an oil should be chosen containing under 2 per cent. impurities and a low percentage of free fatty acids. Lagos oil is best adapted to these requirements. The oil is melted by open steam from a jet introduced through the bung, the melted oil and condensed water running to the store tank through two sieves (about 1/8 inch mesh) to remove the fibrous material and gross impurities. The oil thus obtained contains fine earthy and fibrous material and vegetable albuminous matter which should be removed, as far as possible, since chemicals are wasted in their oxidation and they retard the bleaching. This is best done by boiling the oil for one hour with wet steam and 10 per cent. solution of common salt (2 per cent. dry salt on weight of oil used) in a lead-lined or wooden tank. After settling over night the brine and impurities are removed by running from a cock at the bottom of the vat and the oil is run out into the bleaching tank through an oil cock, situated about seven inches from the bottom.

The bleaching tank is a lead-lined iron tank of the approximate dimensions of 4 feet deep, 4 feet long and 3-1/2 feet wide, holding about 1-1/2 tons. The charge is one ton. A leaden outlet pipe is fixed at the bottom, to which is attached a rubber tube closed by a screw clip. A plug is fitted into the lead outlet pipe from above. Seven inches above the lower outlet is affixed another tap through which the oil is drawn off.

The tank is further equipped with a wet steam coil and a coil arranged to allow thorough air agitation, both coils being of lead. A good arrangement is to use one coil to deliver either air or steam. These coils should extend as nearly as possible over the entire bottom of the tank and have a number of small downward perforations, so as to spread the agitation throughout the mass.

The temperature of the oil is reduced by passing in air to 110° F. and 40 pounds of fine common salt per ton added through a sieve. About one-half of the acid (40 pounds of concentrated commercial hydrochloric acid) is now poured in and this is followed by the sodium bichromate in concentrated solution, previously prepared in a small lead vat or earthen vessel by dissolving 17 pounds of bichromate in 45 pounds commercial hydrochloric acid. This solution should be added slowly and should occupy three hours, the whole mass being thoroughly agitated with air during the addition and for one hour after the last of the bleaching mixture has been introduced. The whole mixture is now allowed to settle for one hour and the exhausted chrome liquors are then run off from the lower pipe to a waste tank. About 40 gallons of water are now run into the bleached oil and the temperature raised by open steam to 150° to 160° F. The mass is then allowed to settle over night.

One such wash is sufficient to remove the spent chrome liquor completely, provided ample time is allowed for settling. A number of washings given successively with short periods of settling do not remove the chrome liquors effectually. The success of the operation depends entirely upon the completeness of settling.

The wash water is drawn off as before and the clear oil run to storage tanks or to the soap kettle through the upper oil cock. The waste liquors are boiled with wet steam and the oil skimmed from the surface, after which the liquors are run out through an oil trap.

By following the above instructions carefully it is possible to bleach one ton of palm oil with 17 pounds of bichromate of soda and 85 pounds hydrochloric acid.

The spent liquors should be a bright green colour. Should they be of a yellow or brownish shade insufficient acid has been allowed and more must be added If low grade oils are being treated more chrome will be necessary, the amount being best judged by conducting the operation as usual and after the addition of the bichromate, removing a sample of the oil, washing the sample and noting the colour of a rapidly cooled sample.

A little practice enables the operator to judge the correspondence between the colour to be removed and the amount of bleaching mixture to be added.

To obtain success with this process the method of working given must be adhered to even in the smallest detail.

AIR BLEACHING OF PALM OIL

The method of conducting this process is identical with the chrome process to the point where the hydrochloric acid is to be added to the oil. In this method no acid or chrome is necessary, as the active bleaching agent is the oxygen of the air.

The equipment is similar to that of the former process, except that a wooden tank in which no iron is exposed will suffice to bleach the oil in. The process depends in rapidity upon the amount of air blown through the oil and its even distribution. Iron should not be present or exposed to the oil during bleaching, as it retards the process considerably.

After the impurities have been removed, as outlined under the chrome process, the temperature of the oil is raised by open steam to boiling. The steam is then shut off and air allowed to blow through the oil until it is completely bleached, the temperature being maintained above 150° F. by occasionally passing in steam. Usually a ton of oil is readily and completely bleached after the air has been passed through it for 18 to 20 hours, provided the oil is thoroughly agitated by a sufficient flow of air.

If the oil has been allowed to settle over night, it is advisable to run off the condensed water and impurities by the lower cock before agitating again the second day.

When the oil has been bleached to the desired colour, which can be determined by removing a sample and cooling, the mass is allowed to settle, the water run off to a waste tank from which any oil carried along may be skimmed off and the supernatant clear oil run to the storage or soap kettle.

In bleaching by this process, while the process consumes more time and is not as efficient in bleaching the lower grade oils, the cost of bleaching is less and with a good oil success is more probable, as there is no possibility of any of the chrome liquors being present in the oil. These give the bleached oil a green tint when the chrome method is improperly conducted and they are not removed.

Instead of blowing the air through it, the heater oil may be brought into contact with the air, either by a paddle wheel arrangement, which, in constantly turning, brings the oil into contact with the air, or by pumping the heated oil into an elevated vessel, pierced with numerous fine holes from which the oil continuously flows back into the vessel from which the oil is pumped. While in these methods air, light and heat act simultaneously in the bleaching of the oil, the equipment required is too cumbersome to be practical.

Investigations in bleaching palm oil by oxygen showed that not only the colouring matter but the oil itself was affected. In bleaching palm oil for 30 hours with air the free fatty acid content rose and titre decreased considerably.

Olive Oil, which comes from the fruit of the olive trees, varies greatly in quality, according to the method by which it is obtained and according to the tree bearing the fruit. Three hundred varieties are known in Italy alone. Since the larger portion of olive oil is used for edible purposes, a lower grade, denatured oil, denatured because of the tariff, is used for soap manufacture in this country. The oil varies in colour from pale green to golden yellow. The percentage of free acid in this oil varies greatly, though the oil does not turn rancid easily. It is used mainly in the manufacture of white castile soap.

Olive oil foots, which is the oil extracted by solvents after the better oil is expressed, finds its use in soap making mostly in textile soaps for washing and dyeing silks and in the production of green castile soaps.

Other oils, as poppy seed oil, sesame oil, cottonseed oil, rape oil, groundnut oil, are used as adulterants for olive oil, also as substitutes in the manufacture of castile soap, since they are cheaper than olive oil.

Cottonseed Oil is largely used in the manufacture of floating and laundry soaps. It may be used for toilet soaps where a white colour is not desired, as yellow spots appear on a finished soap in which it has been used after having been in stock a short time.

Corn Oil and Soya Bean Oil are also used to a slight extent in the manufacture of toilet soaps, although the oils form a soap of very little body.

Corn oil finds its greatest use in the manufacture of soap for washing motor cars. It is further employed for the manufacture of cheap liquid soaps.

Fatty Acids are also used extensively in soap manufacture. While the soap manufacturer prefers to use a neutral oil or fat, since from these the by-product glycerine is obtained, circumstances arise where it is an advantage to use the free fatty acids. Red oil (oleic acid, elaine) and stearic acid are the two fatty acids most generally bought for soap making. In plants using the Twitchell process, which consists in splitting the neutral fats and oils into fatty acids and glycerine by dilute sulphuric acid and producing their final separation by the use of so-called aromatic sulphonic acids, these fatty acids consisting of a mixture of oleic, stearic, palmitic acids, etc., are used directly after having been purified by distillation, the glycerine being obtained from evaporating the wash water.

Oleic acid (red oil) and stearic acid are obtained usually by the saponification of oils, fats and greases by acid, lime or water under pressure or Twitchelling. The fatty acids thus are freed from their combination with glycerine and solidify upon cooling, after which they are separated from the water and pressed at a higher or lower temperature. The oleic acid, being liquid at ordinary temperature, together with some stearic and palmitic acid, is thus pressed out. These latter acids are usually separated by distillation, combined with the press cake further purified and sold as stearic acid.

The red oil, sometimes called saponified red oil, is often semi-solid, resembling a soft tallow, due to the presence of stearic acid. The distilled oils are usually clear, varying in colour from light to a deep brown. Stearic acid, which reaches the trade in slab form, varies in quality from a soft brown, greasy, crumbly solid of unpleasant odour to a snow white, wax-like, hard, odourless mass. The quality of stearic acid is best judged by the melting point, since the presence of any oleic acid lowers this. The melting point of the varieties used in soap manufacture usually ranges from 128° to 132° F. Red oil is used in the manufacture of textile soaps, replacing olive oil foots soap for this purpose, chlorophyll being used to colour the soap green. Stearic acid, being the hard firm fatty acid, may be used in small quantities to give a better grade of soap body and finish. In adding this substance it should always be done in the crutcher, as it will not mix in the kettle. It finds its largest use for soap, however, in the manufacture of shaving soaps and shaving creams, since it produces the non-drying creamy lather so greatly desired for this purpose. Both red oil and stearic acid being fatty acids, readily unite with the alkali carbonates, carbon dioxide being formed in the reaction and this method is extensively used in the formation of soap from them.

RANCIDITY OF OILS AND FATS.

Rancidity in neutral oils and fats is one of the problems the soap manufacturer has to contend with. The mere saying that an oil is rancid is no indication of its being high in free acid. The two terms rancidity and acidity are usually allied. Formerly, the acidity of a fat was looked upon as the direct measure of its rancidity. This idea is still prevalent in practice and cannot be too often stated as incorrect. Fats and oils may be acid, or rancid, or acid and rancid. In an acid fat there has been a hydrolysis of the fat and it has developed a rather high percentage of free acid. A rancid fat is one in which have been developed compounds of an odoriferous nature. An acid and rancid fat is one in which both free acid and organic compounds of the well known disagreeable odours have been produced.

It cannot be definitely stated just how this rancidity takes place, any more than just what are the chemical products causing rancidity. The only conclusion that one may draw is that the fats are first hydrolyzed or split up into glycerine and free fatty acids. This is followed by an oxidation of the products thus formed.

Moisture, air, light, enzymes (organized ferments) and bacteria are all given as causes of rancidity.

It seems very probable that the initial splitting of the fats is caused by enzymes, which are present in the seeds and fruits of the vegetable oils and tissue of animal fats, in the presence of moisture. hold that bacteria or micro-organisms are the cause of this hydrolysis, citing the fact that they have isolated various micro-organisms from various fats and oils. The acceptance of the bacterial action would explain the various methods of preservation of oils and fats by the use of antiseptic preparations. It cannot, however, be accepted as a certainty that bacteria cause the rancidity of fats.

The action of enzymes is a more probable explanation.

The hydrolysis of fats and oils is accelerated when they are allowed to remain for some time in the presence of organic non-fats. Thus, palm oil, lower grades of olive oil, and tallow, which has been in contact with the animal tissue for a long time, all contain other nitrogenous matter and exhibit a larger percentage of free fatty acid than the oils and fats not containing such impurities.

Granting this initial splitting of the fat into free fatty acids and glycerine, this is not a sufficient explanation. The products thus formed must be acted upon by air and light. It is by the action of these agents that there is a further action upon the products, and from this oxidation we ascertain by taste and smell (chemical means are still unable to define rancidity) whether or not a fat is rancid. While some authorities have presumed to isolate some of these products causing rancidity, we can only assume the presence of the various possible compounds produced by the action of air and light which include oxy fatty acids, lactones, alcohols, esters, aldehydes and other products.

The soap manufacturer is interested in rancidity to the extent of the effect upon the finished soap. Rancid fats form darker soaps than fats in the neutral state, and very often carry with them the disagreeable odour of a rancid oil. Further, a rancid fat or oil is usually high in free acid. It is by no means true, however, that rancidity is a measure for acidity, for as has already been pointed out, an oil may be rancid and not high in free acid.

The percentage of free fatty acid is of even greater importance in the soap industry. The amount of glycerine yield is dependent upon the percentage of free fatty acid and is one of the criteria of a good fat or oil for soap stock.

PREVENTION OF RANCIDITY

Since moisture, air, light and enzymes, produced by the presence of organic impurities, are necessary for the rancidity of a fat or oil, the methods of preventing rancidity are given. Complete dryness, complete purification of fats and oils and storage without access of air or light are desirable. Simple as these means may seem, they can only be approximated in practice. The most difficult problem is the removal of the last trace of moisture. Impurities may be lessened very often by the use of greater care. In storing it is well to store in closed barrels or closed iron tanks away from light, as it has been observed that oils and fats in closed receptacles become rancid less rapidly than those in open ones, even though this method of storing is only partially attained. Preservatives are also used, but only in edible products, where their effectiveness is an open question.

CHEMICAL CONSTANTS OF OILS AND FATS

Besides the various physical properties of oils and fats, such as colour, specific gravity, melting point, solubility, etc., they may be distinguished chemically by a number of chemical constants. These are the iodine number, the acetyl value, saponification number, Reichert-Meissl number for volatile acids, Hehner number for insoluble acids. These constants, while they vary somewhat with any particular oil or fat, are more applicable to the edible products and are criteria where any adulteration of fat or oil is suspected.

OIL HARDENING OR HYDROGENATING

It is very well known that oils and fats vary in consistency and hardness, depending upon the glycerides forming same. Olein, a combination of oleic acid and glycerine, as well as oleic acid itself largely forms the liquid portion of oils and fats. Oleic acid ($C_{18}H_{34}O_2$) is an unsaturated acid and differs from stearic acid ($C_{18}H_{36}O_2$), the acid forming the hard firm portion of oils and fats, by containing two atoms of hydrogen less in the molecule. Theoretically it should be a simple matter to introduce two atoms of hydrogen into oleic acid or olein, and by this mere addition convert liquid oleic acid and olein into solid stearic acid and stearine.

For years this was attempted and all attempts to apply the well-known methods of reduction (addition of hydrogen) in organic chemistry, such as treatment with tin and acid, sodium amalgam, etc., were unsuccessful. In recent years, however, it has been discovered that in the presence of a catalyzer, nickel in finely divided form or the oxides of nickel are usually employed, the process of hydrogenating an oil is readily attained upon a practical basis.

The introduction of hardened oils has opened a new source of raw material for the soap manufacturer in that it is now possible to use oils in soap making which were formerly discarded because of their undesirable odours. Thus fish or train oils which had up to the time of oil hydrogenating resisted all attempts of being permanently deodorised, can now be employed very satisfactorily for soap manufacture. A Japanese chemist, Tsujimoto has shown that fish oils contain an unsaturated acid of the composition $C_{18}H_{28}O_2$, for which he proposed the name clupanodonic acid. By the catalytic hardening of train oils this acid passes to stearic acid and the problem of deodourizing these oils is solved.

At first the introduction of hardened oils for soap manufacture met with numerous objections, due to the continual failures of obtaining a satisfactory product by the use of same. Various attempts have now shown that these oils, particularly hardened train oils, produce extraordinarily useful materials for soap making. These replace expensive tallow and other high melting oils. It is of course impossible to employ hardened oils alone, as a soap so hard would thus be obtained that it would be difficultly soluble in water and possess very little lathering quality. By the addition of 20-25% of tallow oil or some other oil forming a soft soap a very suitable soap for household use may be obtained. Ribot discusses this matter fully. Hardened oils readily saponify, may be perfumed without any objections and do not impart any fishy odour to an article washed with same. Meyerheim states that through the use of hydrogenated oils the hardness of soap is extraordinarily raised, so that soap made from hardened cottonseed oil is twelve times as hard as the soap made from ordinary cottonseed oil. This soap is also said to no longer spot yellow upon aging, and as a consequence of its hardness, is able to contain a considerably higher content of rosin through which lathering power and odour may be improved. Hardened oils can easily be used for toilet soap bases, provided they are not added in too great a percentage.

The use of hardened oils is not yet general, but there is little doubt that the introduction of this process goes a long way toward solving the problem of cheaper soap material for the soap making industry.

GREASE

Grease varies so greatly in composition and consistency that it can hardly be classed as a distinctive oil or fat. It is obtained from refuse, bones, hides, etc., and while it contains the same constituents as tallow, the olein content is considerably greater, which causes it to be more liquid in composition. Grease differs in colour from an off-white to a dark brown. The better qualities are employed in the manufacture of laundry and chip soap, while the poorer qualities are only fit for the cheapest of soaps used in scrubbing floors and such purposes. There is usually found in grease a considerable amount of gluey matter, lime and water. The percentage of free fatty acid is generally high.

The darker grades of grease are bleached before being used.

ROSIN

Rosin is the residue which remains after the distillation of turpentine from various species of pines. The chief source of supply is in the United States – Georgia, North and South Carolina. It is a transparent, amber coloured hard pulverisable resin. The better grades are light in colour and known as water white (w. w.) and window glass (w. g.). These are obtained from a tree which has been tapped for the first year. As the same trees are tapped from year to year, the product becomes deeper and darker in colour until it becomes almost black.

The constituents of rosin are chiefly (80-90%) abietic acid or its anhydride together with pinic and sylvic acids. Its specific gravity is 1.07-1.08, melting point about 152.5 C., and it is soluble in alcohol, ether, benzine, carbon disulfide, oils, alkalis and acetic acid. The main use of rosin, outside of the production of varnishes, is in the production of laundry soaps, although a slight percentage acts as a binder and fixative for perfumes in toilet soaps and adds to their detergent properties. Since it is mainly composed of acids, it readily unites with alkaline carbonates, though the saponification is not quite complete and the last portion must be completed through the use of caustic hydrates, unless an excess of 10% carbonate over the theoretical amount is used. A lye of 20° B. is best adapted to the saponification of rosin when caustic hydrates are employed for this purpose, since weak lyes cause frothing. While it is sometimes considered that rosin is an adulterant for soap, this is hardly justifiable, as it adds to the cleansing properties of soap. Soaps containing rosin are of the well-known yellowish colour common to ordinary laundry soaps. The price of rosin has so risen in the last few years that it presents a problem of cost to the soap manufacturer considering the price at which laundry soaps are sold.

ROSIN SAPONIFICATION

As has been stated, rosin may be saponified by the use of alkaline carbonates. On account of the possibility of the soap frothing over, the kettle in which the operation takes place should be set flush with the floor, which ought to be constructed of cement. The kettle itself is an open one with round bottom, equipped with an open steam coil and skimmer pipe, and the open portion is protected by a semi-circular rail. A powerful grid, having a 3-inch mesh, covers one-half of the kettle, the sharp edges protruding upwards.

The staves from the rosin casks are removed at the edge of the kettle, the rosin placed on the grid and beaten through with a hammer to break it up into small pieces.

To saponify a ton of rosin there are required 200 lbs. soda ash, 1,600 lbs. water and 100 lbs. salt. Half the water is run into the kettle, boiled, and then the soda ash and half the salt added. The rosin is now added through the grid and the mixture thoroughly boiled. As carbon dioxide is evolved by the reaction the boiling is continued for one hour to remove any excess of this gas. A portion of the salt is gradually added to grain the soap well and to keep the mass in such condition as to favour the evolution of gas. The remainder of the water is added to close the soap and boiling continued for one or two hours longer. At this point the kettle must be carefully watched or it will boil over through the further escape of carbon dioxide being hindered. The mass, being in a frothy condition, will rapidly settle by controlling the flow of steam. The remaining salt is then scattered in and the soap allowed to settle for two hours or longer. The lyes are then drained off the top. If the rosin soap is required for toilet soaps, it is grained a second time. The soap is now boiled with the water caused by the condensation of the steam, which changes it to a half grained soap suitable for pumping. A soap thus made contains free soda ash 0.15% or less, free rosin about 15%. The mass is then pumped to the kettle containing the soap to which it is to be added at the proper stage. The time consumed in thus saponifying rosin is about five hours.

NAPHTHENIC ACIDS

The naphtha or crude petroleum of the various provinces in Europe, as Russia, Galacia, Alsace and Rumania yield a series of bodies of acid character upon refining which are designated under the general name of naphthenic acids. These acids are retained in solution in the alkaline lyes during the distillation of the naphtha in the form of alkaline naphthenates. Upon adding dilute sulphuric acid to these lyes the naphthenates are decomposed and the naphthenic acids float to the surface in an oily layer of characteristic disagreeable odour and varying from yellow to brown in colour. In Russia particularly large quantities of these acids are employed in the manufacture of soap.

The soaps formed from naphthenic acids have recently been investigated and found to resemble the soaps made from coconut oil and palm kernel oil, in that they are difficult to salt out and dissociate very slightly with water. The latter property makes them valuable in textile industries when a mild soap is required as a detergent, e. g., in the silk industry. These soaps also possess a high solvent power for mineral oils and emulsify very readily. The mean molecular weight of naphthenic acids themselves is very near that of the fatty acids contained in coconut oil, and like those of coconut oil a portion of the separated acids are volatile with steam. The iodine number indicates a small content of unsaturated acids.

Naphthenic acids are a valuable soap material but except in Russia the soap is not manufactured to any extent.

ALKALIS

The common alkali metals which enter into the formation of soap are sodium and potassium. The hydroxides of these metals are usually used, except in the so called carbonate saponification of free fatty acids in which case sodium and potassium carbonate are used.

A water solution of the caustic alkalis is known as lye, and it is as lyes of various strengths that they are added to oils and fats to form soap. The density or weight of a lye is considerably greater than that of water, depending upon the amount of alkali dissolved, and its weight is usually determined by a hydrometer. This instrument is graduated by a standardized scale, and while all hydrometers should read alike in a liquid of known specific gravity, this is generally not the case, so that it is advisable to check a new hydrometer for accurate work against one of known accuracy. The Baumé scale has been adopted in the United Sates while in the United Kingdom a different graduation known as the Twaddle scale is used. The strength of a lye or any solution is determined by the distance the instrument sinks into the solution, and we speak of the strength of a solution as so many degrees Baumé or Twaddle which are read to the point where the meniscus of the lye comes on the graduated scale. Hydrometers are graduated differently for liquids of different weights. In the testing of lyes one which is graduated from 0° to 50° B. is usually employed.

Caustic soda is received by the consumer in iron drums weighing approximately 700 lbs. each. The various grades are designated as 60, 70, 74, 76 and 77%. These percentages refer to the percentage of sodium oxide (Na_2O) in 100 parts of pure caustic soda formed by the combination of 77-1/2 parts of sodium oxide and 22-1/2 parts of water, 77-1/2% being chemically pure caustic soda. There are generally impurities present in commercial caustic soda. These consist of sodium carbonate, sodium chloride or common salt and sometimes lime. It is manufactured by treating sodium carbonate in an iron vessel with calcium hydroxide or slaked lime, or by electrolysis of common salt. The latter process has yet been unable to compete with the former in price. Formerly all the caustic soda used in soap making was imported, and it was only through the American manufacturer using a similar container to that used by foreign manufacturers that they were able to introduce their product. This prejudice has now been entirely overcome and most of the caustic soda used in this country is manufactured here.

CAUSTIC POTASH

The output of the salts containing potassium is controlled almost entirely by Germany. Formerly the chief source of supply of potassium compounds was from the burned ashes of plants, but the salt mines of Stassfurt, Saxony, Germany, were discovered. The salt there mined contains, besides the chlorides and sulphates of sodium, magnesium, calcium and other salts, considerable quantities of potassium chloride, and the Stassfurt mines at present are practically the entire source of all potassium compounds, in spite of the fact that other localities have been sought to produce these compounds on a commercial basis, especially by the United States government.

After separating the potassium chloride from the magnesium chloride and other substances found in Stassfurt salts the methods of manufacture of caustic potash are identical to those of caustic soda. In this case, however, domestic electrolytic caustic potash may be purchased cheaper than the imported product and it gives results equal to those obtained by the use of the imported article, although not all soap makers agree.

SODIUM CARBONATE (SODA ASH)

While carbonate of soda is widely distributed in nature the source of supply is entirely dependent upon the manufactured product. Its uses are many, but it is especially important to the soap industry in the so called carbonate saponification of free fatty acids, as a constituent of soap powders, in the neutralization of glycerine lyes and as a filler for laundry soaps.

The old French Le Blanc soda process, which consists in treating common salt with sulphuric acid and reducing the sodium sulphate (salt cake) thus formed with carbon in the form of charcoal or coke to sodium sulphide, which when treated with calcium carbonate yields a mixture of calcium sulphide and sodium carbonate (black ash) from which the carbonate is dissolved by water, has been replaced by the more recent Solvay ammonia soda process. Even though there is a considerable loss of salt and the by-product calcium chloride produced by this process is only partially used up as a drying agent, and for refrigerating purposes, the Le Blanc process cannot compete with the Solvay process, so that the time is not far distant when the former will be considered a chemical curiosity. In the Solvay method of manufacture sodium chloride (common salt) and ammonium bicarbonate are mixed in solution. Double decomposition occurs with the formation of ammonium chloride and sodium bicarbonate. The latter salt is comparatively difficultly soluble in water and crystallizes out, the ammonium chloride remaining in solution. When the sodium bicarbonate is heated it yields sodium carbonate, carbon dioxide and water; the carbon dioxide is passed into ammonia which is set free from the ammonium chloride obtained as above by treatment with lime (calcium oxide) calcium chloride being the by-product.

Sal soda or washing soda is obtained by recrystallizing a solution of soda ash in water. Large crystals of sal soda containing but 37% sodium carbonate are formed.

POTASSIUM CARBONATE

Potassium carbonate is not extensively used in the manufacture of soap. It may be used in the forming of soft soaps by uniting it with free fatty acids. The methods of manufacture are the same as for sodium carbonate, although a much larger quantity of potassium carbonate than carbonate of soda is obtained from burned plant ashes. Purified potassium carbonate is known as pearl ash.

Only one other point needs to be made. This is that in soap manufacture it is best to use soft water, but where it is necessary to use hard water, a number of softeners can be applied to stop the result being an inert type of soap filled with lime and thus of little use domestically or industrially.

www.ingramcontent.com/pod-product-compliance
Lightning Source LLC
Chambersburg PA
CBHW032021190326
41520CB00007B/565